Quantum Physics for Babies

This is a ball.

This ball has energy.

This is a ball.

This ball has zero energy.

All balls are made of atoms.

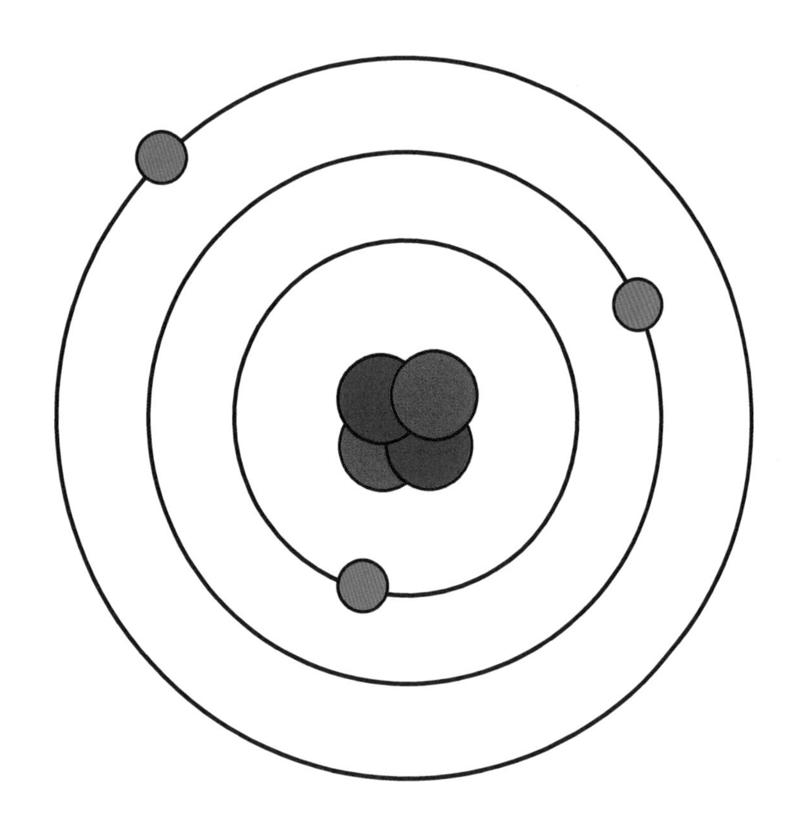

There are neutrons.

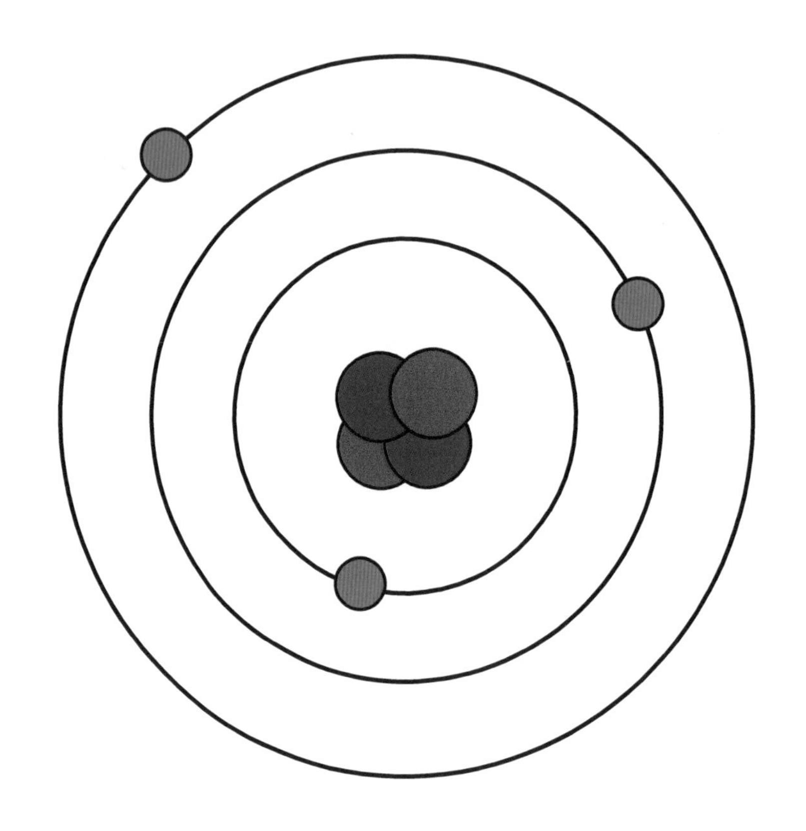

And protons.

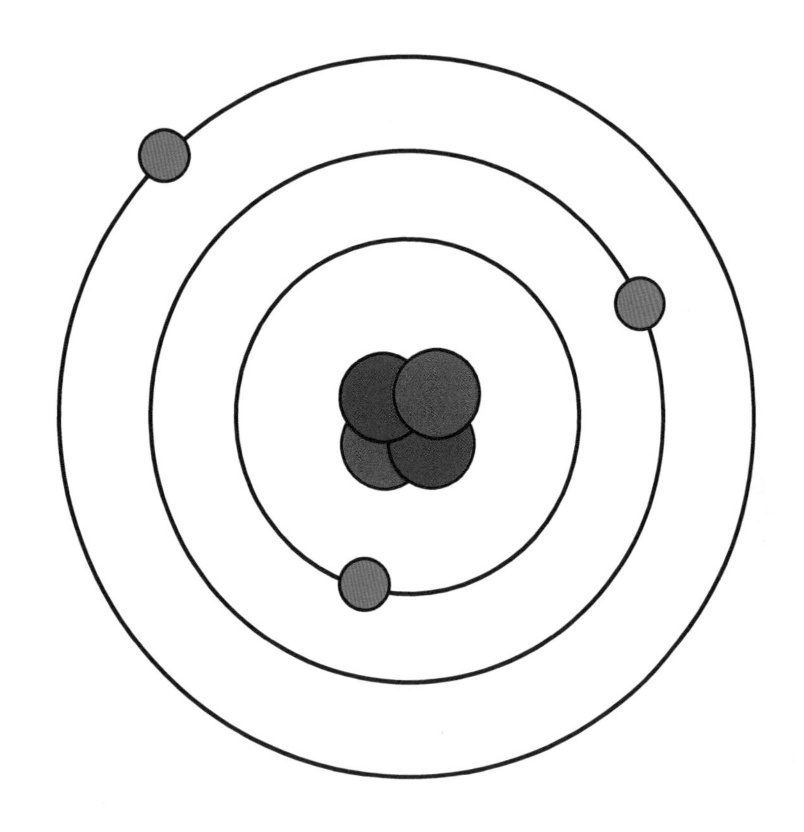

And electrons.

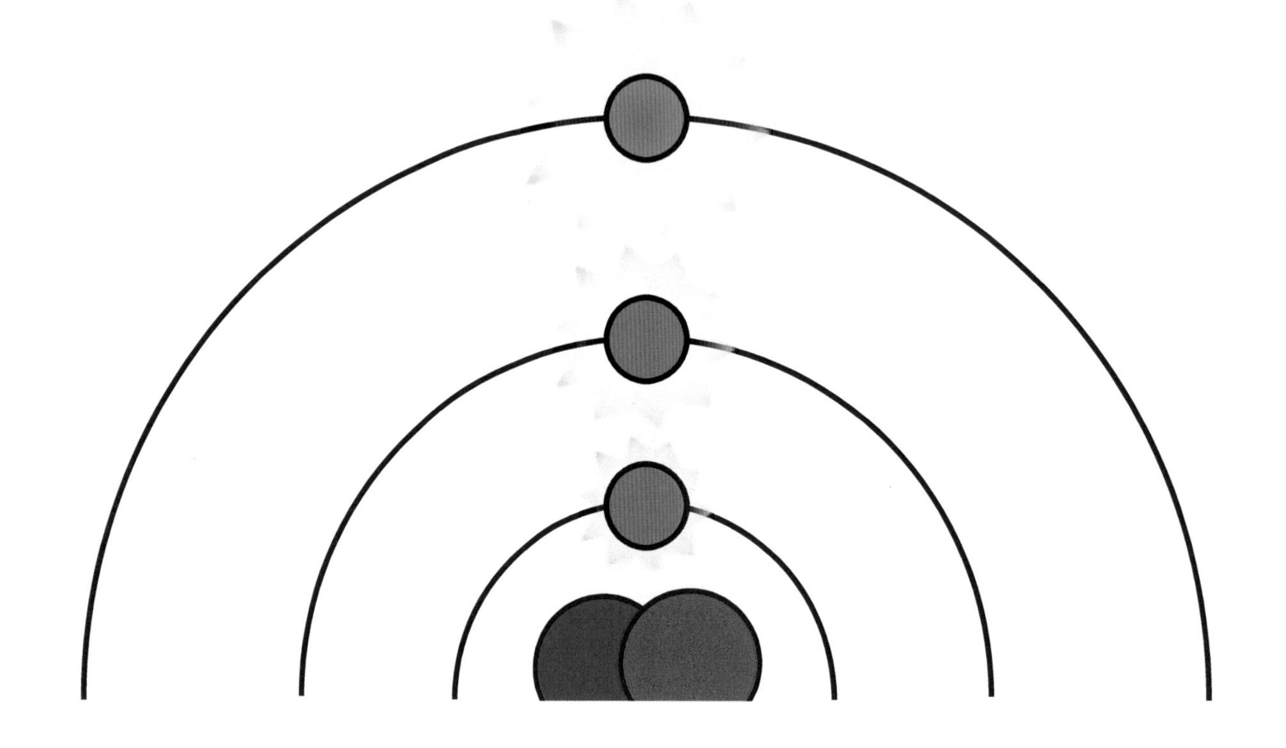

Electrons have energy.

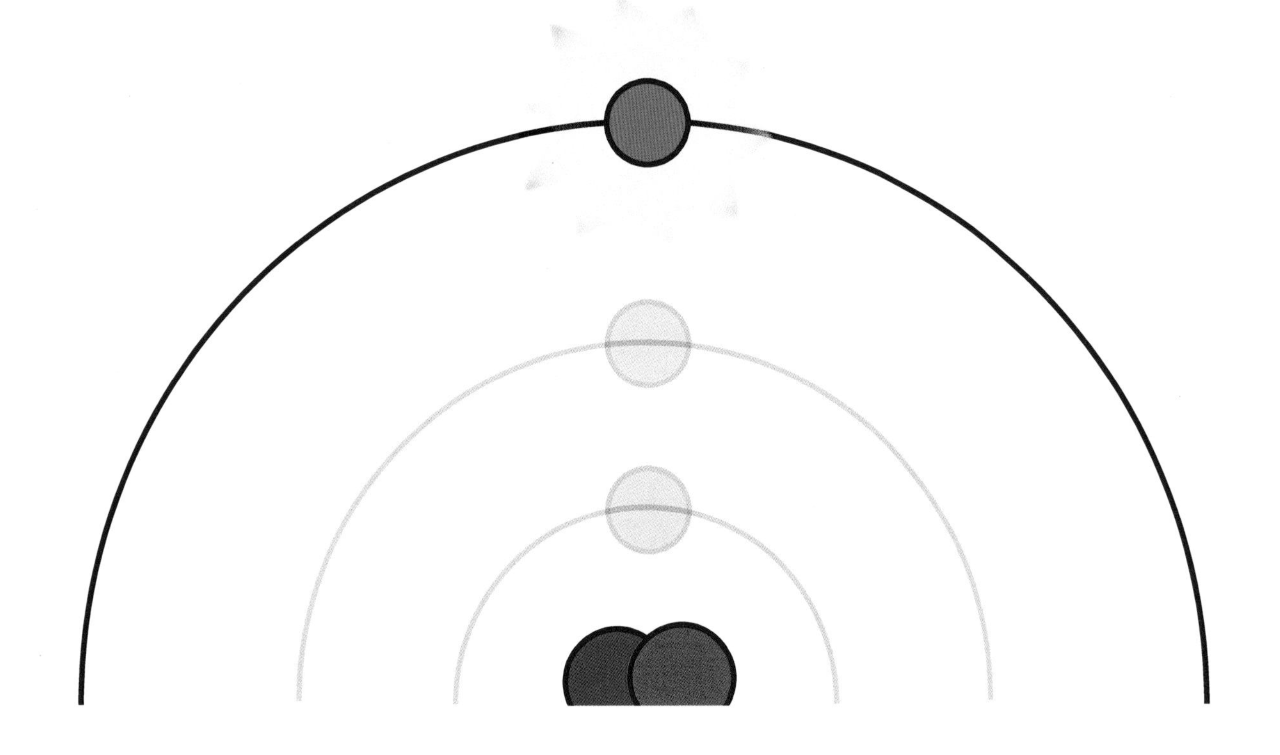

This electron has the most energy.

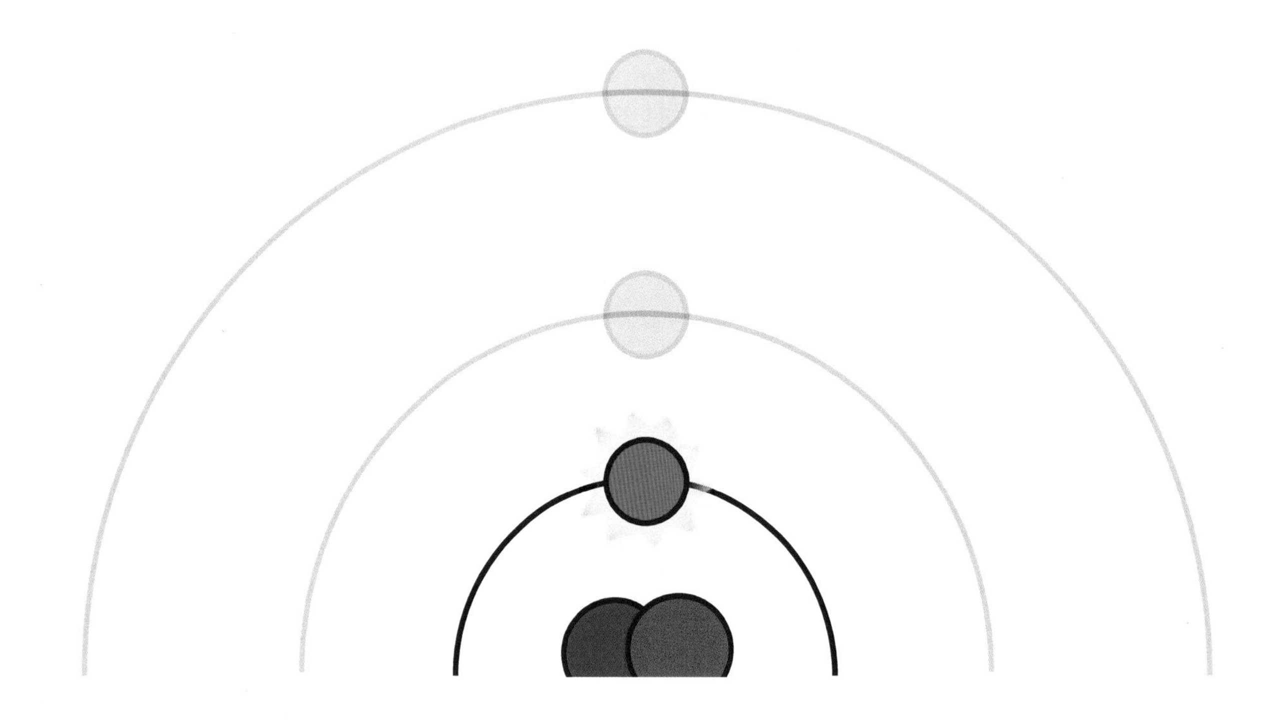

This electron has the least energy.

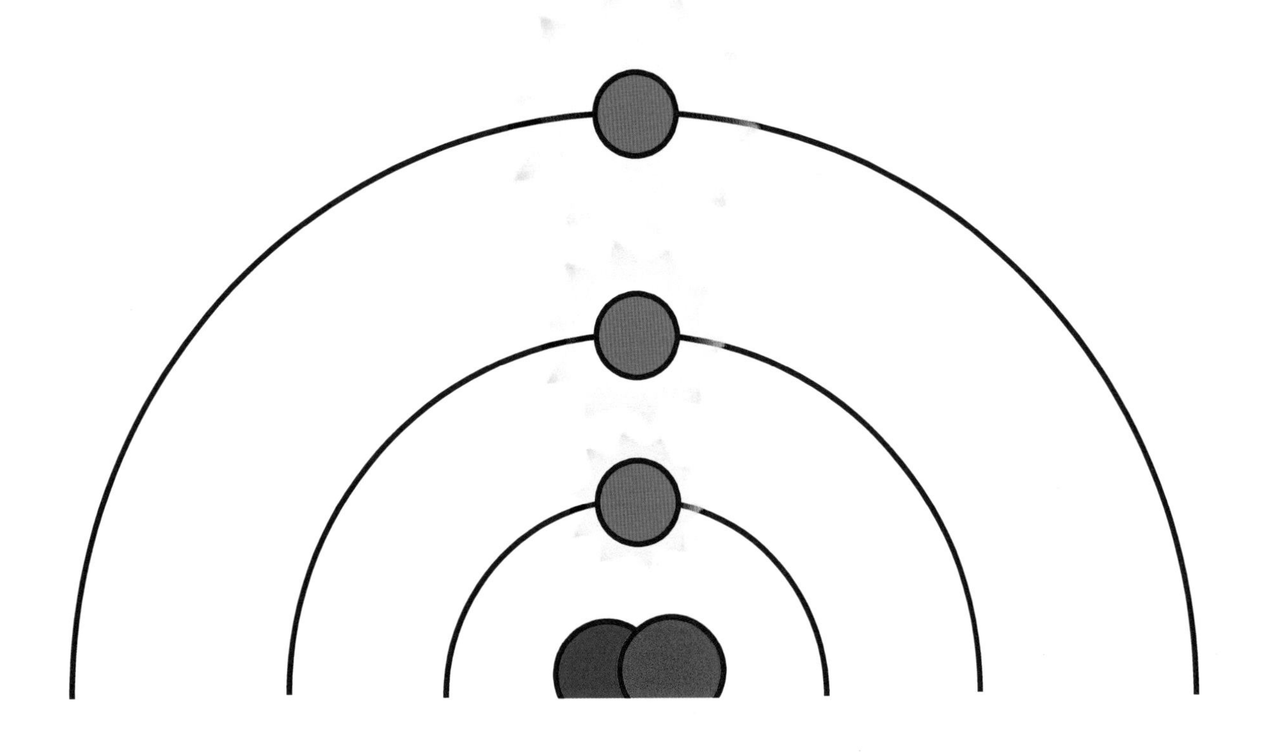

Energy **is** quantized.

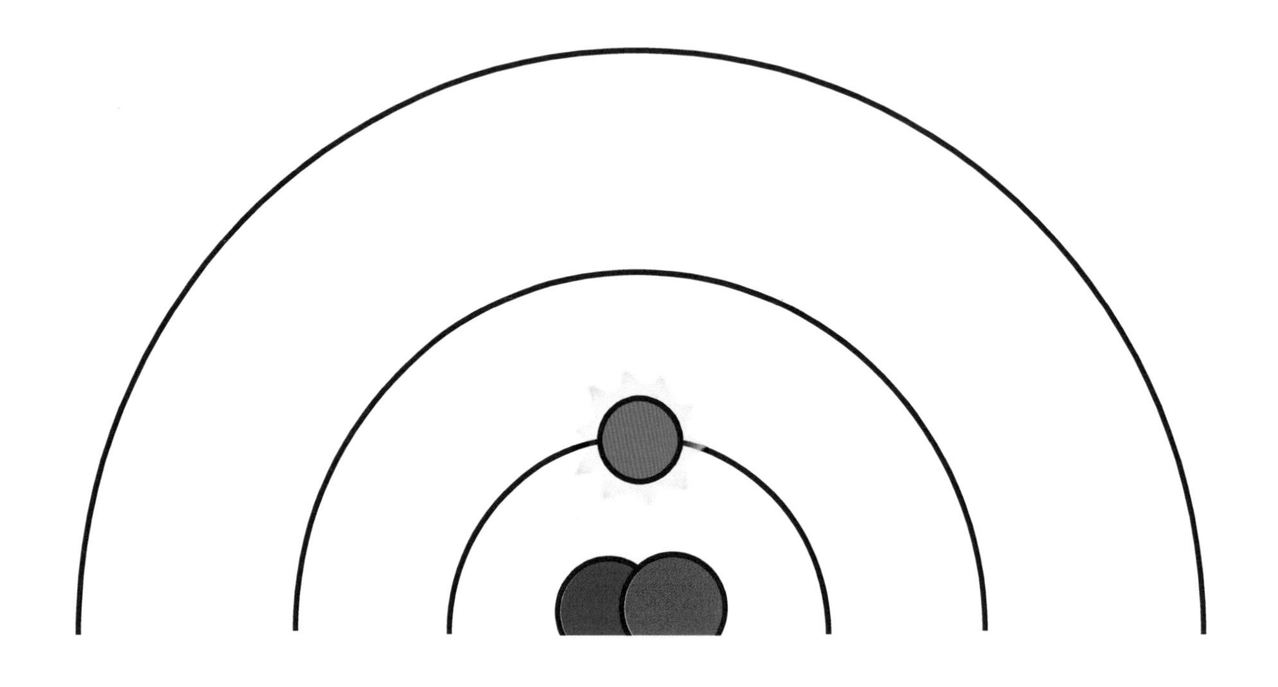

An electron can be here.

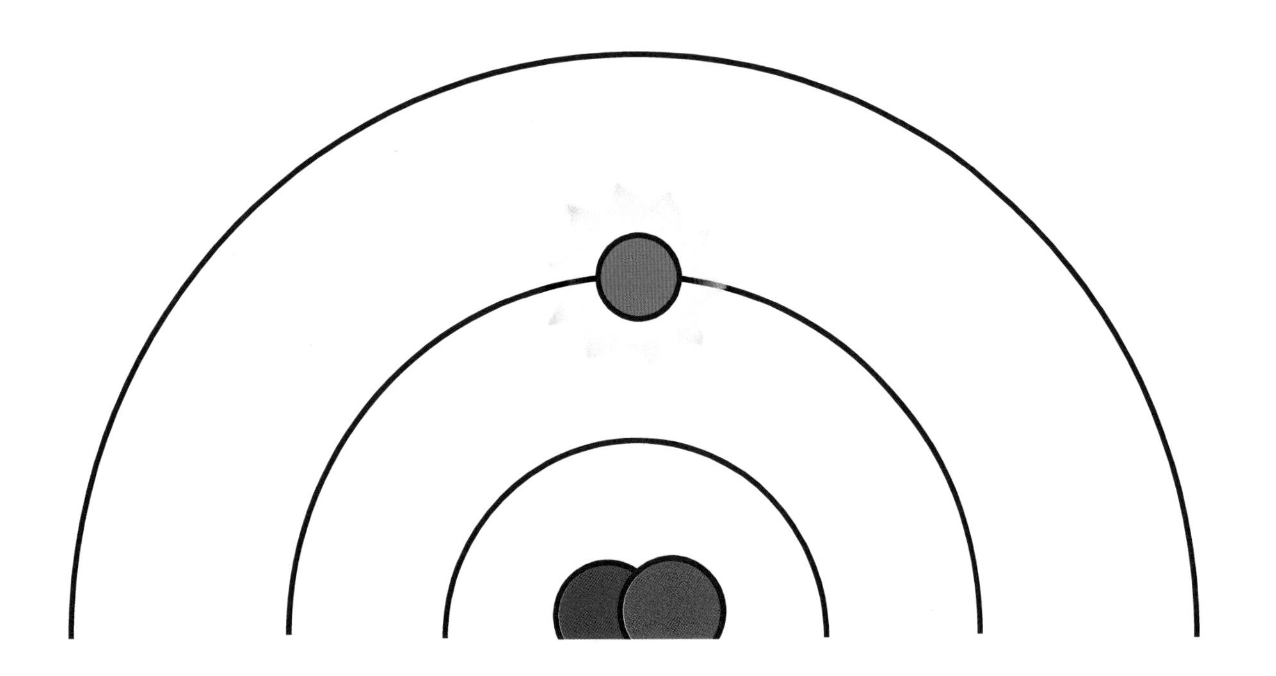

Or here.

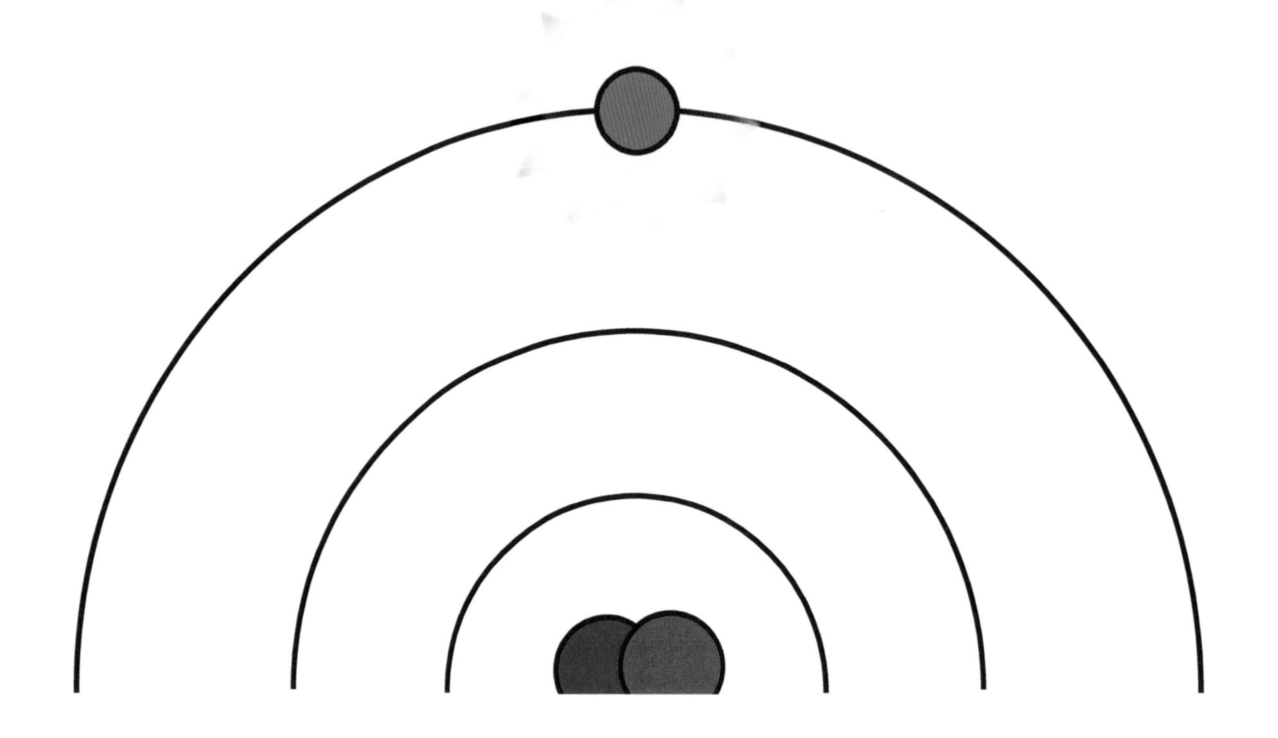

Or here.

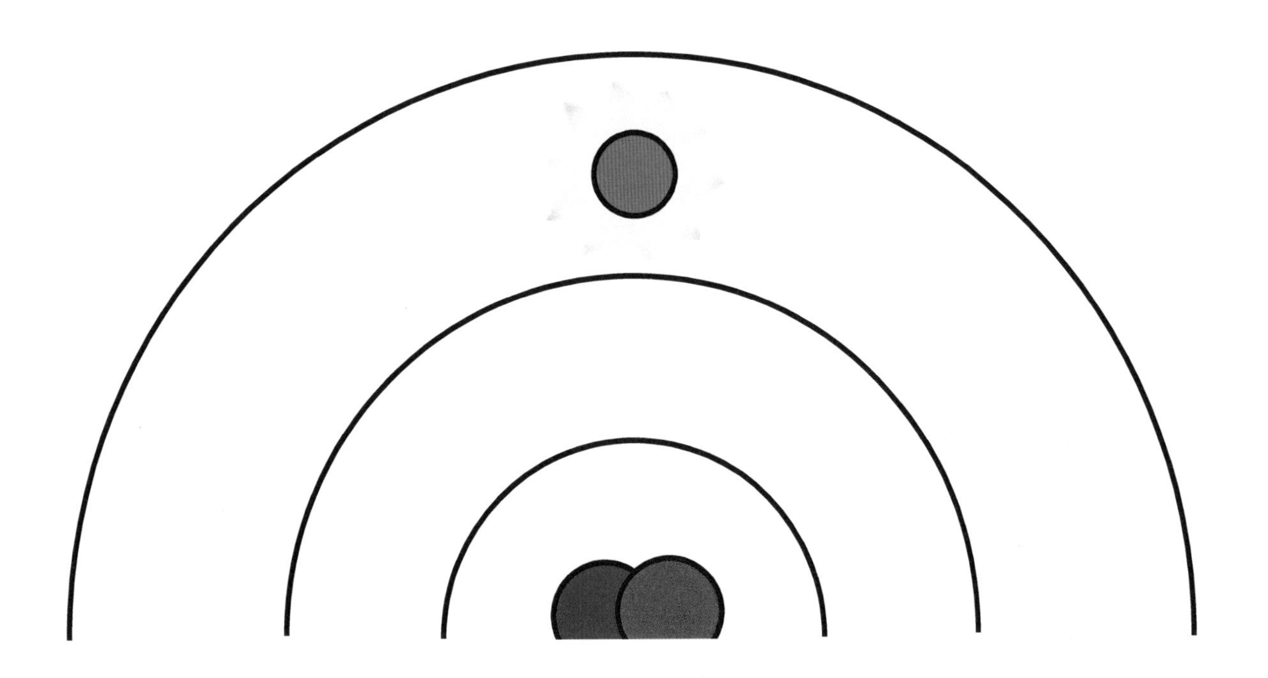

But not here.

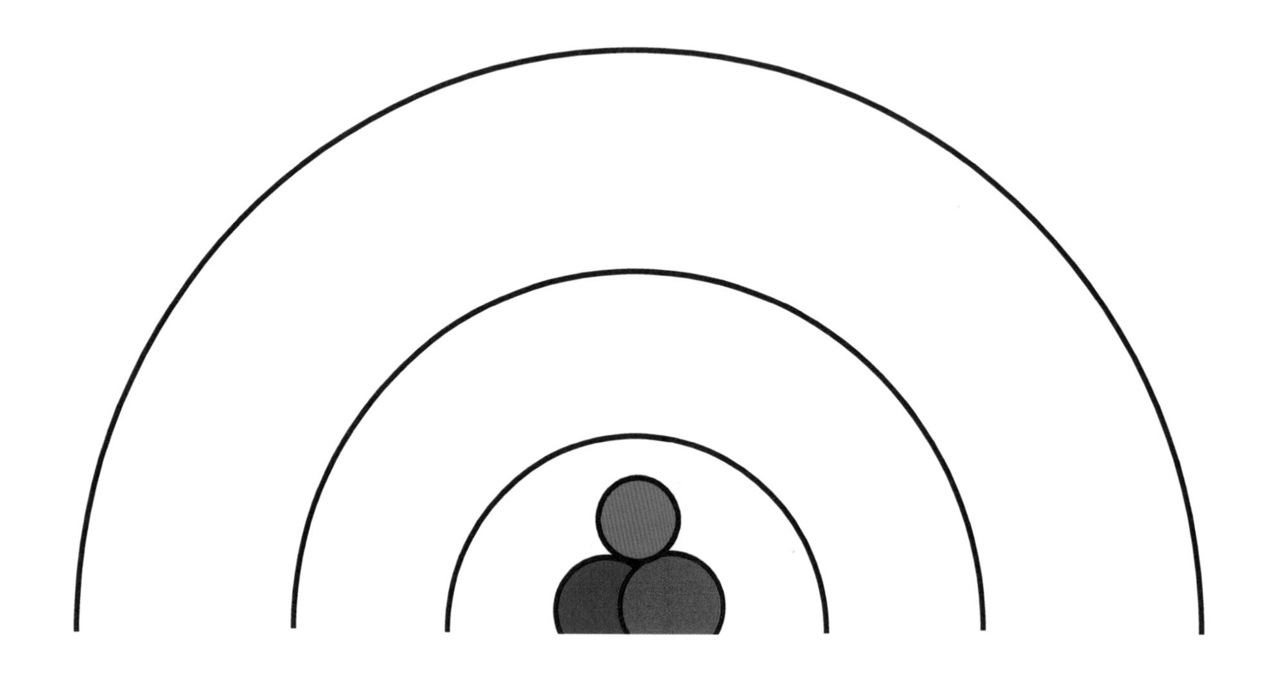

Or here.

There are no electrons with zero energy.

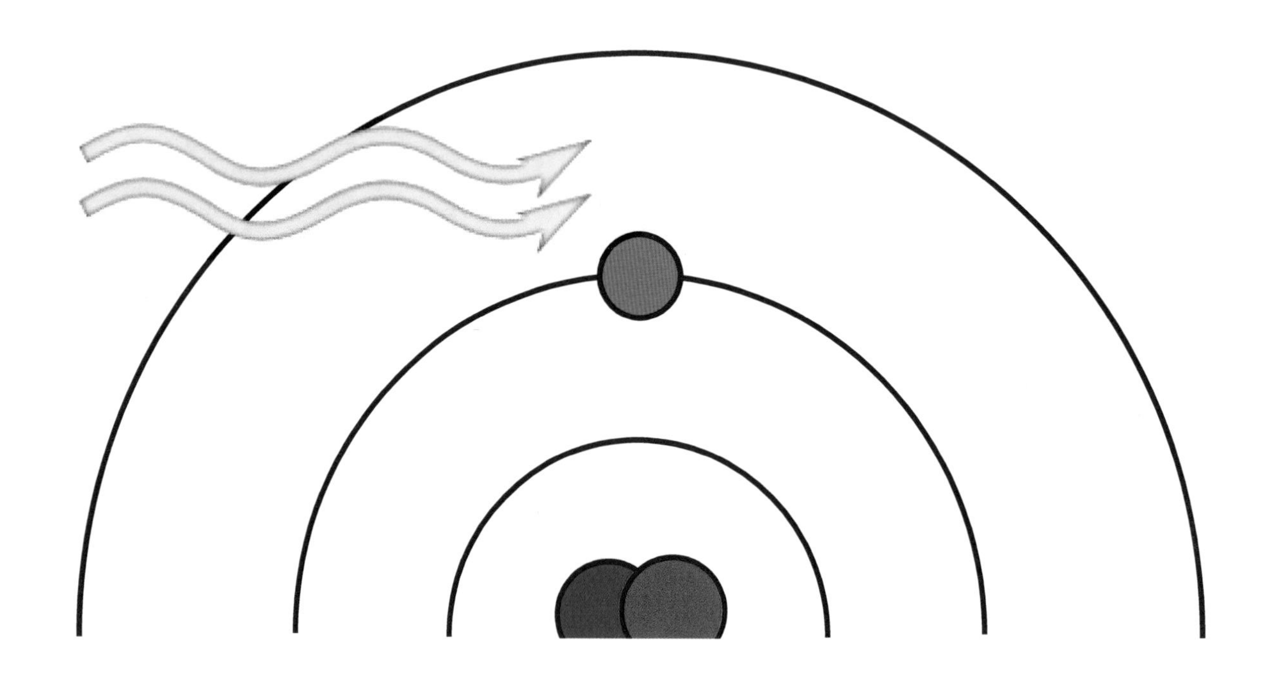

An electron can take energy.

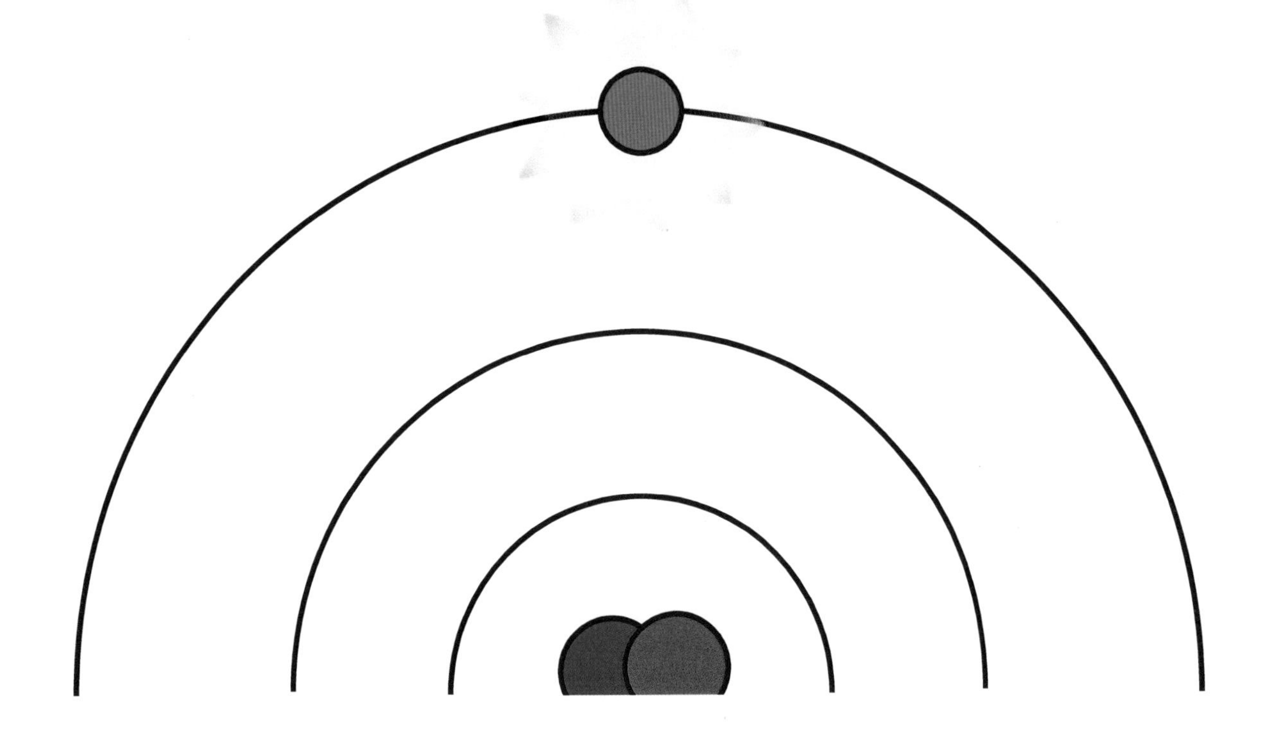

To jump up.

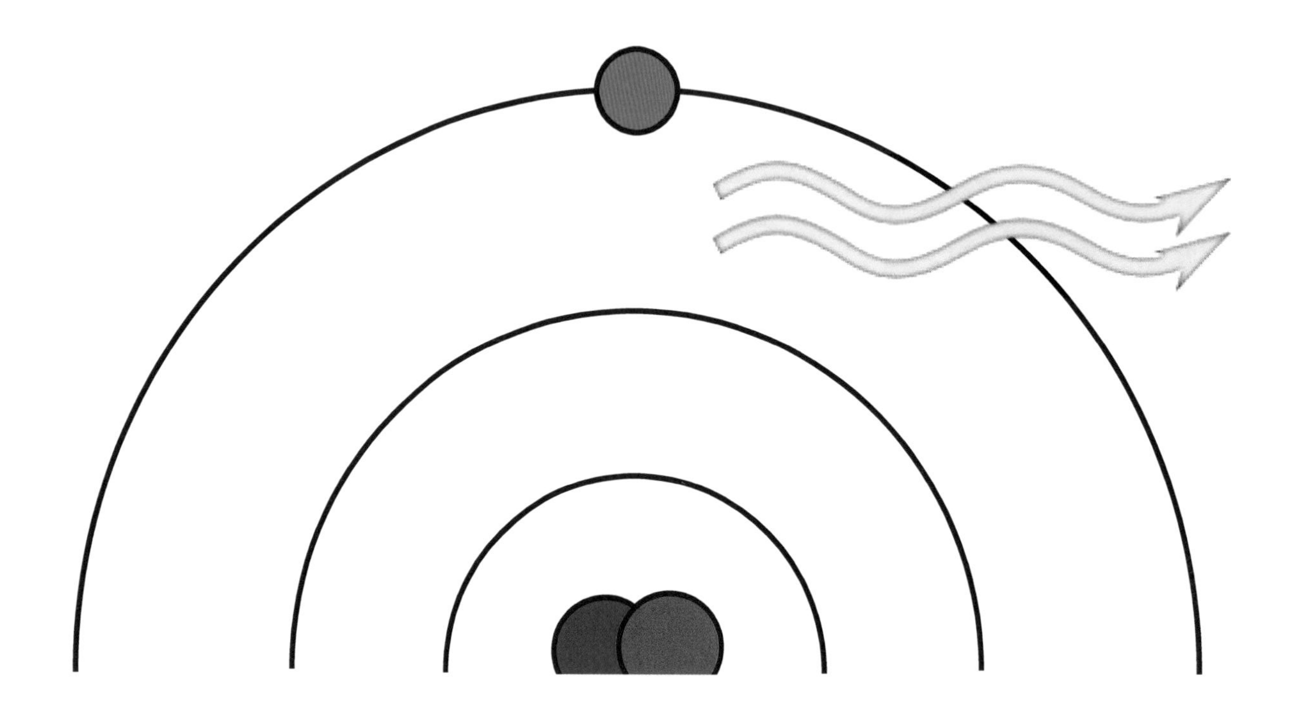

And must give
energy.

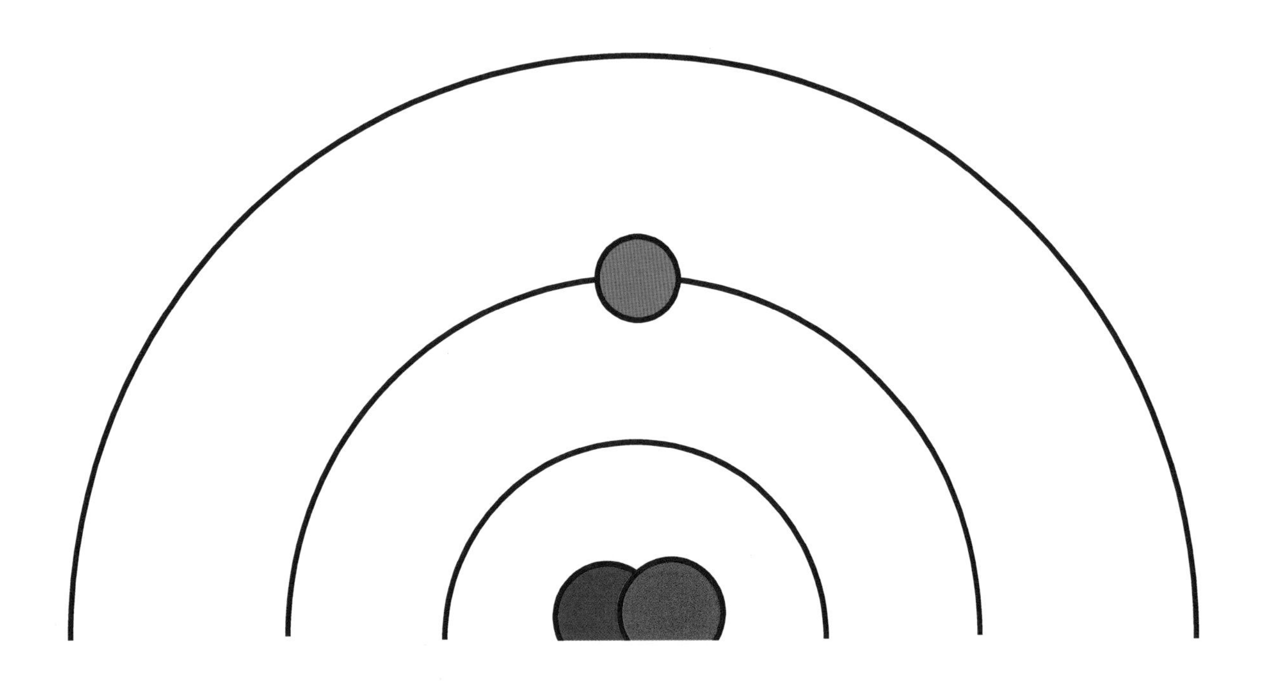

To fall down.

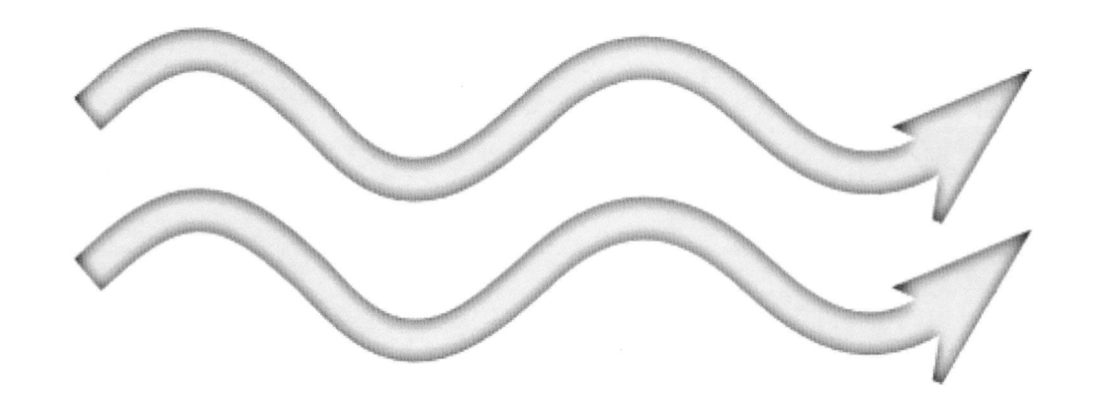

This amount of

energy is a quantum.

Now you are a quantum physicist.

Chris Ferrie is a physicist, mathematician and father of three budding young scientists. He obtained his doctorate in Mathematical Physics from the University of Waterloo in Waterloo, Canada and currently holds a postdoctoral fellowship at the University of Sydney in Sydney, Australia.

Chris believes it is never too early to introduce children to the wild and wonderful world of physics!

Made in the USA
Lexington, KY
22 July 2015